BRITISH BIRDS EGGS AND NESTS

Where, when and how to find and recognise them.

REDSHANK

by ERIC POCHIN

President of the Leicestershire & Rutland Ornithological Society.

Drawings by JOHN READ
from the actual eggs

INDEX OF BRITISH BIRDS' EGGS REFERRED TO IN THIS BOOK

The CUCKOO makes no nest but lays her eggs in the nests of other species, the commoner foster-parents being the Hedge-sparrow, Robin, Redstart, the pipits, wagtails, finches, warblers and chats. Cuckoo's eggs vary much in colour, but seem to follow roughly the pattern of those of the foster-bird. Indeed it is now believed that cuckoos, at any rate the hens, are divisable into types according to the foster-parent that both rears them and on whom they foist their own eggs.

CUCKOO

The REED-WARBLER builds a most beautiful nest of grass, reed-flowers, etc, with a deep cup lined with finer grass, feathers, wool, hair and reed-flowers. It is usually sewn onto three or four reed stems growing out of the water but sometimes onto rank vegetation at the water's edge.

REED-WARBLER

FOREWORD

In this little book it is not possible to include all species of eggs, nor more than one example of those chosen; each is typical and allowance must be made both in size and colouration.

The nests and eggs are not grouped in species, but under a series of headings, covering the localities where they are usually to be found.

Birds are conservative creatures and each species builds its nest in the same sort of place and the same type of nest. It must be remembered, however, that birds such as the Robin, Jackdaw and Tits are not very particular and may choose a variety of sites. In such cases they are included in the site-group where the nest is usual and other sites are mentioned in the text.

SWAN

The SWAN makes a bulky nest of reeds and water plants near the margin of water and lays her 5-12 greenish-white eggs in a hollow at the top. Laying begins in April, there is one brood, and each egg is about 4½ x 3 ins, the largest British egg.

NESTS IN TREES

ROOKERY

Members of the crow family, such as the ROOK, CARRION-CROW and MAGPIE, all nest in the tree-tops. The Jackdaw, (page 31) does so sometimes, choosing an old nest in a rookery, but more often builds in hollow trees, old buildings or cliffs.

ROOKS nest together in a rookery but the CARRION-CROW builds alone. Both use twigs broken from living trees, with a lining of moss, grass and wool.

Ravens (page 30) and Hooded-crows (page 22) sometimes choose trees, but more usually a cliff and occasionally the open moor.

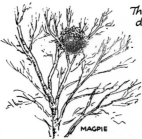

The MAGPIE makes a domed nest of thorns which is a very good protection for both eggs and young. The cup is reinforced with earth and lined with roots, etc. Often the bird will build in a high hedge in preference to a tree.

MAGPIE

COLOUR OF EGGS

The illustrations in this book are of typical examples, but there is a great variability in the eggs of each species. Thus those that are usually heavily marked may sometimes be almost plain, while another that is normally lightly freckled may be heavily spotted

The SISKIN, which is of the finch family and a northern species, builds a nest similar to that of the Linnet (page 12) but in the higher branches of fir-trees. Other finches, such as the Greenfinch (page 11), Hawfinch (page 11) and Redpoll (page 12) will at times choose a tree in preference to a hedge or bush. The nest of the Siskin is made of twigs, grass, moss, lined with wool, hair and feathers.

The CROSSBILL, also of the finch family, prefers to nest in the tops of fir-trees. First it lays a platform of twigs; then upon this foundation it makes a nest of dry grass, wool, moss and roots, with a lining of finer materials.

The SPOTTED-FLYCATCHER chooses a depression in the trunk of a tree or the angle made by a branch, but it will also build its nest on an ivy-covered wall; in fact it is not particular, any cranny in a tree or wall will suit. There it makes a not very tidy nest of grass and hair, lining it with similar soft materials.

SPOTTED FLYCATCHER

The MISTLE-THRUSH often chooses the fork of a tree or sometimes the angle where two branches meet. It is rather an untidy, bulky affair of grass and moss, with large pieces of sheep's wool often sticking out. It has a mud cup, lined with grass.

The GOLDCREST is a very clever little nest builder, constructing its home like a hammock slung under the branch of a yew or fir-tree, generally about six feet from the ground. It is beautifully made of moss, lichen and spider's web, lined with feathers.

GOLDCREST

COLOUR OF EGGS

After the egg has been formed in the oviduct of the hen bird and during the growth of the shell, the ground colour is deposited. The spots or markings are added later in successive applications just prior to laying.

NESTS IN TREES

SPARROW-
HAWK

The SPARROW-HAWK builds on the top of an old nest, such as that of a Crow, Magpie or Wood-pigeon, in a tall tree; or it may merely patch one up with twigs and bits of bark. It is usually bulky and rather flat.

The KESTREL is lazy so far as nest building is concerned. It lays its eggs in any old nest that is suitable. Sometimes the ledge of a cliff is chosen or a cranny in a barn or a ruin.

The BUZZARD makes a large nest of sticks, heather and roots, lining it with grass and twigs; and then finally decorating it with fresh green leaves, which are often added to as the eggs are laid. Like the Kestrel, this bird also often builds on the ledge of a cliff. Another smaller falcon, the Hobby, which is comparatively rare, like the Kestrel uses an old nest in which to lay its eggs. In appearance the latter are rather like those of the Merlin (page 23).

COLOUR OF EGGS

The colour effect of all eggs that are marked or speckled tends to break up the otherwise plain ground, to destroy the oval contour, in short, to camouflage the egg. Some eggs have so little marking that it is wholly insufficient to give them protection, but this does not disprove the general principle.

Birds which hide their eggs in holes, such as the Kingfisher, Owls and Sand-martin, lay pure white eggs, while those that are laid in the open, such as Gulls' and Plovers' are the most heavily marked or spotted. It would seem, therefore, that egg marking is a tendency, not yet fully developed, towards camouflage.

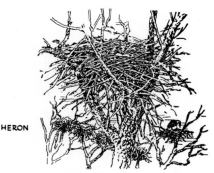

HERON

HERONS, although birds of the waterside, build their nests in the tree-tops in colonies called heronries. Each nest is quite a bulky structure, made of sticks, lined with twigs and grass, and as in a rookery, several may be placed in one tree. Occasionally, if no trees are available, nests may be found among reeds on the ground.

The LONG-EARED OWL is no nest builder, but lays its eggs in the old nest of a Crow, Hawk or Wood-pigeon. It chooses a deep thick fir-wood where there is plenty of shade and where it is not likely to be disturbed. Usually the old nest is flattened out into a kind of platform before the eggs are laid.

The WOOD-PIGEON and TURTLE-DOVE build very similar nests, which, although only a platform of sticks and twigs, are so interwoven that they are remarkably secure. Sometimes it is possible to see the eggs from underneath through the nest itself. These nests may also be found in high hedges, especially that of the Turtle-dove. The latter begins to lay at the end of May and has one or two broods and the eggs are much smaller.

WOOD-PIGEON

※※※※※※ COLOUR OF EGGS ※※※※※※

A glance through this book will give some idea of the range of colours that are met with, and although in the various families of birds this seems to follow some sort of pattern, the reason for such a wide range is unknown.

NESTS IN HOLES IN TREES

TREE-
SPARROW

The *TREE-SPARROW* likes best a hole in an old tree for its nest, but it will sometimes build in a nest-box or hole in a wall. The nest is made of straw, grass, roots and wool, amply lined with feathers. Often it will add fresh leaves of nearby plants to the lining. In a suitable place several pairs will often nest together in a sort of colony.

The *TREE-CREEPER* finds the place where a piece of bark has become dislodged on the trunk of a tree, or a crevice, and there builds a nest of moss and grass lined with wool and feathers. Sometimes the nest is made in the cranny of a wall or among ivy stems.

The *NUTHATCH* is a proper builder, for it uses mud to fashion the entrance to its nest in a hole in a tree-trunk. This it does so that the entrance is only just large enough for it to squeeze in. It never pecks out the hole itself. The nest inside is made of leaves grass and wood-chips.

NUTHATCH

In some species both male and female build the nest, while in others one or other does all the work. Similarly in the brooding of the eggs, and the tending of the young, the work may be shared or carried out entirely by one bird.

The GREAT TIT will make
its nest in almost any hole, in
a tree or wall; in a nest-box,
a flower-pot or an old can.
Often the nest is so deep that
it is remarkable that the
young birds are able to climb
out. Moss, grass and wool are
used as the foundation, with a
lining of hair

GREAT TIT

The BLUE TIT is very similar
to the last named in its nesting
habits. Providing the entrance-hole is not too large any
crevice or cranny is equally suitable. It is remarkable
how small the aperture may be. The materials used
are similar to those in the Great-tit's nest.

The MARSH-TIT and COAL-TIT prefer a hole in a tree
or stump. The nest in each case is made of moss, lined
with a thick pad of felted hair and fur so that unless
the bird is seen the actual species cannot be verified.
All tits have the habit of covering their eggs with nest
material until the clutch is complete.

Unlike the Spotted-flycatcher, the PIED-FLYCATCHER
selects a hole in an old tree, building its nest inside.
Woodpeckers' holes are sometimes used. It has been
known to nest in
a wall. Leaves,
grass and moss
are employed
in construction,
with a lining
of hair.

PIED-FLYCATCHER

Remember that you can study birds' eggs very much better
in the nest than as dusty and faded specimens in a glass case.
Join your local naturalist society and send in all the informa-
tion you can obtain about birds' nests. The kind of thing that
will be valuable is ·
    The average number of eggs in the clutches of the
    commoner species.
    The earliest and latest dates of nesting.
    The exact incubation periods of commoner birds.
    Complete nesting cycles from the building of the nest to the
    departure of the young birds.
    Anything special about the nesting habits of an individual
    bird.

NESTS IN HOLES IN TREES

GREEN
WOODPECKER

All the woodpeckers bore out a hole in a tree trunk to make a nest. The GREEN WOODPECKER makes the largest hole, about 2½ inches diameter. The GREAT SPOTTED WOODPECKER comes next in size and the LESSER SPOTTED WOODPECKER the smallest 1¾ inches diameter. The glossy eggs are laid on the wood chippings at the bottom and may be 12 inches or more below the entrance. The average size of the eggs is as follows: Green Woodpecker, 1¼ ins. long. Great Spotted Woodpecker, 1 inch long. Lesser Spotted Woodpecker, ¾ inch long.

The WRYNECK does not bore out its own nest-hole, but finds one already made. As with the woodpeckers, no actual nest is built inside, the eggs resting on the bottom of the cavity. The bird is now rare, being found only in the wooded districts of south-eastern England.

The TAWNY OWL prefers a hollow tree if it can find one, but sometimes the eggs are laid in an old building or even in a rabbit-burrow. No actual nest is made, the eggs resting on the floor of the place that has been chosen. The LITTLE OWL is similar in its nesting habits, but the eggs are smaller, 4 to 6 in number, and the bird begins to lay somewhat later.

TEXTURE OF EGGS

Egg shells vary very much in thickness, usually according to the size of the egg, but not entirely so. Generally speaking, eggs that are laid on the open ground or in a scrape-nest have the stronger shells, especially those of some of the sea-birds

NESTS IN BUSHES and HEDGES

JAY

The JAY is a member of the crow family, but builds its rather bulky nest of twigs, lined with some finer materials, in the undergrowth of woods, not as a rule very high.

The GREENFINCH also makes a somewhat large nest for the size of the bird. It is constructed of twigs moss and roots, lined with finer material; roots, hair and feathers. The Hawfinch builds a similar type of nest but without moss and feathers. Both birds are fond of orchards and gardens.

The GOLDFINCH'S nest is one of the most beautiful of all British bird's. It is built of bents, roots, moss and lichen, woven together with wool and lined with down. It is extremely neat and small. Goldfinches are also fond of orchards and gardens.

Another clever builder is the CHAFFINCH. The nest is very like that of the Goldfinch, but is lined with hair and feathers and often decorated with scraps of birch-bark and paper, while spider's web is used to sew it together.

CHAFFINCH

The largest British bird's egg is the Mute Swan's, which is about 4½ inches long; that of the Golden Eagle is 3 inches. The smallest, the Goldcrest's, is only ½ inch long.

NESTS IN BUSHES and HEDGES

LINNET

The LINNET is a finch, and its nest conforms to the type, though it is usually found in the lower hedge-rows and bushes; it is very fond of gorse. Several pairs will often build in close proximity. Nests and eggs of the finches are much alike, and the birds should be seen before identification is certain. The Lesser Redpoll's nest has a twiggy foundation, is rougher and is lined with white down, hair and feathers. The eggs are Linnet-like, but smaller.

The BULLFINCH makes a nest of fine twigs as a platform with a little moss, and lines the cup with blackish roots.

The LONG-TAILED TIT is another master-builder in the bird world, for it constructs a domed nest of moss, lichen and hair, felted together with cobweb, and then lines the inside with thousands of feathers. The entrance is on one side near the top and the sitting bird's tail is folded over its back.

LONG-TAILED TIT

The RED-BACKED SHRIKE likes a thick hawthorn bush or a bramble patch in which to make its nest of green moss and grass, lined with hair, roots and feathers. It often returns to the same nesting-place year after year.

The BLACKCAP builds a nest that is typical of the larger warblers both as regards situation and construction. Somewhat loosely made of bents, roots and grass, and lined with finer materials and hair, it is usually to be found in hedges and bushes near the

BLACKCAP

ground. The Garden-warbler generally builds lower than the Blackcap and the nest is more substantial with moss and leaves added. The eggs are like those of the Blackcap.

The WHITETHROAT is another warbler, but the nest has a very deep cup that is lined with dark or black hair, bits of wool and down. Sometimes it is placed almost on the ground. The Lesser Whitethroat prefers very thick bushes and hedges and the nest, usually placed higher up, is smaller and less substantial than the larger bird's. The eggs are smaller and more heavily marked.

The nest of the SONG-THRUSH needs little description; it is made of grass and roots and lined with a cup of mud. That of the BLACKBIRD is similar but the

mud cup is again lined with finer grasses. Both birds build in similar situations and are not very particular, haystacks, woodstacks and buildings often being chosen.

SONG-THRUSH

The HEDGE-SPARROW likes a thick, low hedge for its nest which it builds of moss, twigs and bents, and lines with moss, wool and hair. There is no more lovely picture than this bird's nest with its bright blue eggs.

⊢◇◇◇◇◇◇◇◇◇◇◇◇◇◇◇◇◇◇◇◇◇◇◇⊣

TEXTURE OF EGGS

Some eggs are smooth or even polished, the Kingfisher's and Woodpeckers' for instance. Some, such as those of the larger Gulls, are grained, others are very rough, while the Cormorants' and Grebes' are covered with a thick layer of chalk over the shell. There is no explanation for this.

NESTS IN UNDERGROWTH
in or near woods.

WILLOW-
-WARBLER

Being a close relation of the Lark, the TREE-PIPIT makes its nest on the ground among undergrowth or in a tussock, using dry grass and moss with a lining of finer grass and hair. Both nest and eggs are very like those of the Meadow-pipit (page 20) and Rock-pipit (page 26) and care must be used to obtain a correct identification. The Rock-pipit will only be found near the shore.

The Leaf-warblers: Chiffchaff, WILLOW-WARBLER and WOOD-WARBLER, all make a domed nest on or near the ground of dead leaves, grass and moss. The first two use feathers for a lining, while the Wood-warbler uses hair and fine bents only. The eggs of all three are much alike, but the last named eggs are usually more heavily marked and the Chiffchaff's tend to violet markings. Another warbler that may be met with is the Grasshopper-warbler, which builds a typical warbler nest of grass and dead leaves, lined with fine bents and hair. The eggs, similar in size to the Whitethroat's, are cream, thickly spotted with red-brown, and normally six in number.

The NIGHTINGALE builds its nest in a tangle of dense undergrowth. It is an untidy affair of dead leaves and grass, and lined with finer materials of the same nature. Sometimes the eggs are very finely mottled.

NIGHTINGALE

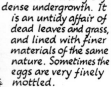

NESTS on the GROUND
in or near woods

NIGHTJAR

The NIGHTJAR makes no nest at all, but lays its two eggs on the ground among last year's dead leaves and bracken in the open woodland. Their marbled marking, however, makes them rather difficult to see.

The WOODCOCK makes a simple nest typical of the wader family using a few leaves and moss to line a hollow among the bracken or heather.

Four eggs are the general rule, their points all being placed to the centre, as is the case of the Snipe.

WOODCOCK

The PHEASANT also makes a scanty nest, in a scrape lined with dead leaves and grass under a bush, bramble or similar cover. The hen sits very closely, relying on the camouflage of her plumage to protect her.

SIZE AND SHAPE OF EGGS

Eggs are roundish or elongated and pointed at one or both ends; the actual shape varying with and within the species. Those laid in the normal cup-nest are usually the ordinary egg shape; those laid in holes tend to be rounder; those in open ground nests are elongated, except for the Plovers' and Waders', which are very pointed at one end so that they may be laid points together in the centre.

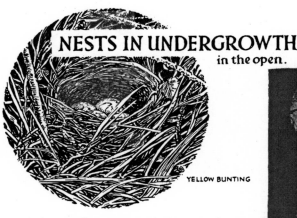

NESTS IN UNDERGROWTH
in the open.

YELLOW BUNTING

Rough lanes and commons are the favourite places for the nest of the Buntings, of which three species may be met with in this country, besides the Reed-bunting that prefers the waterside (page 24). The nests are very similar; the CORN-BUNTING uses bents and grass, lining it with fine grass, roots and hair; the YELLOW BUNTING, grass and moss with a similar lining; while the CIRL BUNTING uses a large amount of moss. The latter's eggs are like those of the Yellow Bunting, but smaller and more boldly marked.

The WHINCHAT prefers a low bush or a tussock of herbage in which to build. The nest is made of dry grass and moss, lined with fine bents and hair. The Stonechat makes a similar nest but often uses feathers in the lining. The latter's eggs, 5-6 in number, are paler and greener, thickly freckled with fine red spots.

The PARTRIDGE hides her nest in a nettle patch or among brambles and bracken, and lines the scrape with grass and dead leaves. When she leaves the nest she covers the eggs over with leaves.

The Red-legged Partridge nests in a like manner but the eggs are larger and spotted with red-brown.

PARTRIDGE

NESTS IN BANKS & WALLS

PIED
WAGTAIL

PIED WAGTAIL

Almost any hole a few feet from the ground will suit the PIED-WAGTAIL as a nest-site. Leaves, twigs, moss and roots are used in its construction with a lining of grass, hair and feathers.

The WHEATEAR likes to build its nest in a wall or a pile of rocks on the open moorland. Sometimes it will choose a rabbit-hole or similar situation and several nests may often be found in close proximity. Grass with a lining of hair, wool and fur are used in its construction.

The REDSTART is rather like a Robin in its choice of a nesting place but generally selects a more open situation. The nest itself is made of roots and grass, lined with feathers and hair.

REDSTART

〰〰〰〰〰〰〰〰〰〰〰〰〰〰〰〰〰〰〰〰〰

SIZE AND SHAPE OF EGGS

The reason why Plovers' and Waders' eggs are pointed and laid points to the centre is because they are large for the size of the bird, and laid in this way they occupy the smallest possible area the bird must cover in brooding them.

Most ground birds lay extra large eggs so that the chicks, when hatched, are strong enough to run about and take care of themselves at once.

NESTS IN BANKS & WALLS

ROBIN

The ROBIN likes a cosier place for its nest than the Redstart; a ditch or a mossy bank with plenty of cover is a favourite spot. Leaves, hair, grass and moss are used in its construction, with a lining of finer materials.

The nest of the WREN is domed and is carefully fitted into any recess that will accommodate it, but a bank is a very favoured place. The materials used match the surroundings as far as is possible, and may be moss, leaves, lichen, dead bracken or grass, and for the lining, feathers.

WREN

Look for a sandy bank for nests of the SAND-MARTIN. There are sure to be more than one, for these birds nest in colonies. In the face of a steep bank they scratch out holes, two or three feet long, and at the end form a round cavity which they line with bits of straw and feathers. Sometimes there may be fifty nests or more in a short stretch of bank.

<----------- 2/3 ft. ----------->

SAND-MARTIN

NESTS IN BANKS NEAR STREAMS

GREY WAGTAIL

The GREY-WAGTAIL likes to find a rocky ledge on the bank of a swift-flowing mountain stream for its nest. It is usually well hidden and some feet above the water. Grass, twigs and moss are used in the main construction with a lining of hair and an odd feather.

The DIPPER'S nest is domed like the Wren's but larger. It is always placed near a swift rocky stream in some crevice in the bank or rocks, under a bridge or similar situation. It is made of moss and grass, with a lining of dead leaves.

DIPPER

The steep earth bank of a river or stream is the place to find the hole of a KINGFISHER'S nest. The bird bores it out two or three feet long and at the end makes a small round chamber. Fish disgorged by the bird form a little mat for the eggs.

2' to 3'

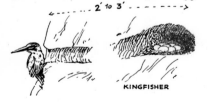

KINGFISHER

NESTS IN OPEN FIELDS

SKYLARK

The SKYLARK likes the open meadowlands and chooses a tuft of grass or a horse's hoof-mark in which to build a nest of dry grass, lined with finer materials. Sometimes roots and hair are used.

The WOOD-LARK prefers a little scrub and a few trees near its nest. This is built in a depression or in the shelter of some herbage, of grass and moss lined with fine grass and a little hair. The bird is very local.

The nest of the MEADOW-PIPIT will be found on the rough pastures and upland meadows of the higher ground. It is well hidden among the growing herbage and is made of grass, lined with finer grass and hair. Sometimes it is very like that of the Tree-pipit (page 14) and the Rock-pipit (page 26).

The YELLOW-WAGTAIL also likes an open situation in meadowland but is not particular providing there is plenty of cover. Growing crops are much favoured, i.e. roots and potatoes The nest, placed in or under a tuft, is made of grass and hair, lined with finer material.

YELLOW-WAGTAIL

It is always difficult to find a plover's nest because it is usually in the middle of a bare field or on the open moor; and the bird is very shy. The eggs, too, are very like their surroundings.

The LAPWING, which is often called the Green Plover, is no exception as regards its nest, but it prefers meadows and ploughed fields in the more lowland districts.

Both nest and eggs are very like those of the Golden Plover.

LAPWING

The GOLDEN PLOVER nests on rough hill pastures and moorland, often among burnt heather. The nest is a mere scrape, usually very scantily lined with a few bents and pieces of heather. Plover's eggs, like many of the waders, are more or less pointed at one end, and in the nest the points are all toward the centre.

The CORNCRAKE is quite different from the last named; it chooses deep grassland in which to hide the simple nest that is only a hollow in the ground, lined with grass and similar material. Unfortunately the bird often makes its nest in mowing grass, so that the eggs are likely to be destroyed by the mowing-machine.

CORNCRAKE

><><><><><><><><><><><><><><><><><><><><><

SIZE AND SHAPE OF EGGS

The eggs in this book are all drawn to natural, average size, and it will be seen how varied they are, but this does not altogether depend on the size of the bird that lays them. The Cuckoo lays a very small egg because it must go into the small nest of the foster-parent. The Guillemot lays a large egg so that the chick, as soon as it is hatched, is strong enough to withstand the rough weather of an open sea cliff.

NESTS ON MOUNTAIN AND MOORLAND

HOODED-
CROW

Open moorland, cliffs or trees may be selected by the HOODED-CROW for its nest-site. Sticks, roots, grass and sea-weed are used in the actual structure, with a lining of fur, wool, hair and grass. It is very like that of the Raven (page 30) and Carrion-crow (page 4).

The TWITE, a cousin of the Linnet (page 12) likes the open moorland, where it makes a nest in the heather and rough grass, or in a crevice of rocks on the hill-side. Roots, grass, twigs and moss are used, with a lining of hair and wool, much the same as that of the Linnet.

The RING-OUZEL builds a nest similar to that of the Blackbird (page 13) except that it is always on the open moor among heather and rocks, and the materials used are mostly heather and bracken.

Wasteland, moor, marsh or sand-dunes are the haunt of the SHORT-EARED OWL, and the nest, such as it is, (though it is the only owl that attempts to build one) is generally made among rough grass and heather. Actually the eggs are laid in a hollow of the ground with little or no materials. The egg is almost identical with that of the Long-eared Owl (page 7) but 4-8 are laid about the end of April and one brood is usual.

⁘⁘⁘ NUMBERS OF EGGS ⁘⁘⁘

A clutch of eggs varies according to the species from one to about fifteen, but the reason for this is unknown, except that birds do not normally increase or decrease in a given area, so the larger the clutch of eggs and number of young reared, the greater must the mortality also be in that particular species.

The MERLIN'S nest is a mere scrape in the ground among the heather, though sometimes an old nest in a tree may be used or the ledge of a cliff. A scanty lining of twigs, heather and grass may be added but the eggs generally rest on the ground.

MERLIN

The CURLEW'S nest will also be found among heather or rough grass on the open hills or upland pastures. It is made in a slight hollow and is lined with grass and bents.

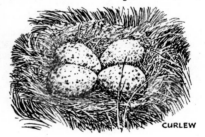

CURLEW

The Stone-curlew nests in the southern part of England on downs, open heaths and arable land. No material is used, the two eggs being laid on the bare ground. They are usually buff, blotched, spotted and streaked with brown and ash-grey, and are slightly smaller than those of the Curlew.

A scrape among the heather or moorland grass lined with grass and feathers serves as a nest for the RED GROUSE. The Ptarmigan's nest and eggs are very similar but the bird prefers the higher mountains, the nest is found above 2000 feet altitude. The Black-grouse prefers the edges of pine plantations on the mountain sides and the eggs are larger and less heavily marked

〰〰〰〰〰 NUMBERS OF EGGS 〰〰〰〰〰

It has been found that the numbers of eggs per clutch is larger in the north than in the south. This might be accounted for by the longer hours of daylight and consequent larger food supply

NESTS at the MARGIN of WATER

REED-BUNTING

The REED-BUNTING usually builds in low herbage near the ground, somewhere in the vicinity of water. The nest is made of grass, lined with hair and fine grass. A little moss is used sometimes.

Dense herbage and undergrowth, generally near water, is the favourite place for the SEDGE-WARBLER'S nest. Moss and grass are used, with a lining of hair, willow-down and feathers. It is not so deep, nor so carefully woven into the reed-stems as the Reed-warbler's (page 3)

The eggs of the DUCK family are not portrayed in this book; except that of the Sheld-duck, which is larger than the average. All Ducks that build near fresh water make very similar nests in a hollow of the ground, hidden by long grass, rushes, bracken and low undergrowth. The materials used include grass, leaves and moss, with a lining of the bird's own down. The number of eggs, their size and colour, varies considerably, as the table on page 25 shows. One brood only is reared; the Mallard commences to lay in March, and the rest in May.

The LITTLE GREBE makes an island of water-weeds for a nest at the edge of a lake or pond with a slight hollow in the centre for the eggs, which it covers when leaving the nest. The nest of the Great Crested Grebe and the bird's habits are similar, but the eggs are larger. In both cases the latter soon become stained very dirty brown by contact with the weed.

LITTLE GREBE

The SNIPE likes marshy ground with coarse grass and rushes in which to hide its nest. It is usually placed under or in a tussock and lined with grass to form a cup. It is well hidden and the bird sits closely.

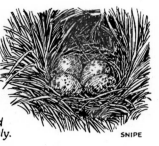

SNIPE

The REDSHANK builds a similar nest in the same sort of surroundings (see title page) but likes a little more cover. The egg is rather like the Golden Plover (page 21) but a little smaller (½" shorter). Four eggs, and one brood only.

Edges of rivers, streams and lakes are the favourite places for a COMMON SANDPIPER'S nest among herbage or cover of some kind, though occasionally in the open. The nest is only a hollow lined with grass.

The nest of the WATERHEN is made of grass, rushes, water-plants and leaves, and may be situated in rushes, herbage or bushes at the edge of ponds, rivers, and streams. Sometimes it is well hidden, but often is quite exposed. The nesting habits of the Water-rail are similar, but the eggs are very much smaller. The Coot builds a nest very like the Waterhen but it is usually in the shallows at the water-edge. The eggs are larger and stone-coloured,

WATERHEN

spotted all over with small purplish-brown spots 7-9 are usual, laying begins in March and 2-3 broods are reared.

EGGS OF THE DUCK FAMILY

TEAL	- - -	8-10 eggs	about 1⅜ in. long.	Greenish/buff
MALLARD	10-12	„	2¼ in. „	Greenish/buff-grey
SHOVELER	- 8-12	„	2 in. „	Greenish/buff
PINTAIL	- - 7-9	„	2⅛ in.	Yellowish/Green-cream
POCHARD	- 6-11	„	2⅜ in. „	Greenish-grey
TUFTED DUCK	8-12	„	2⅜ in. „	Greenish-grey

NESTS on the SHORE and ISLANDS

OYSTER-CATCHER

The OYSTER-CATCHER is not very particular where it places its nest, providing it is near to the sea. Sometimes on the sands or rough ground; sometimes on a rocky cliff or among heather and grass in the dunes. a shallow scrape is lined with bents or pieces of sea-weed, shells and pebbles, and there the eggs are laid.

The nest of the ROCK-PIPIT is very like that of the Meadow and Tree-pipit, the material used being dry grass with hair added to the grass lining. It is usually situated in a crevice of the rocks or among grass and heather, and is well hidden.

The SHELD-DUCK is a sea-duck. but lays its eggs in a rabbit-hole or one it has made itself, often 3 or 4 feet long. Several pairs may nest in close proximity. The actual nest is lined with the bird's down, grass and moss.

The EIDER, another sea-duck. makes a nest of grass and sea-weed lined with her own down. (hence the word eider-down') among rocks and rough herbage never far from the sea-edge. As in the case of the Sheld-duck. several nests may be found near one another The eggs, 4-6 in number. are pale olive to buff in colour and about 3 inches long. Laying begins in May and only 1 brood is reared.

EIDER

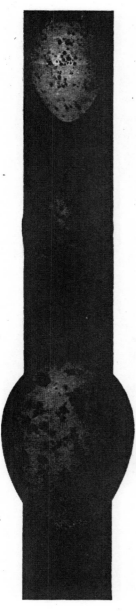

The RINGED-PLOVER nests on sand or pebble beaches, usually in the open without any attempt at cover. No actual nest is made except for a few pieces of broken shell or bents. The four eggs are always laid with their points turned inwards.

The TERNS, of which there are several species, generally nest in colonies on the shore. The eggs are laid, with very little by way of nesting material on the sand, pebbles or rock. The eggs are all much alike and two or three in number. The Sandwich-tern's are about 2 inches long and the Little-tern's, 1¼ inches long.

Several species of GULLS may be met with but most have similar nesting habits and eggs, though the latter vary considerably in size and colouration. The one illustrated is typical, while that of the Herring-gull is about 2¾ inches long; the Lesser Black-backed Gull's 2¾ inches, and the Great Black-backed Gull's 3 inches long. All lay two or three eggs in May and only one brood is reared. Gulls usually nest in colonies; on islands, rough cliffs or the shore and the nest is but a heap of grass, seaweed, feathers and heather.

The Black-headed Gull is somewhat different from those mentioned above because it often nests in huge colonies far from the sea but always near water. The eggs, about 2¼ inches long, are similar to the others but vary much in colouration.

GULL

Clutches vary even with individual birds of a species; thus one Song Thrush may lay four eggs, but another, five. One Waterhen may lay seven eggs, and another, ten. There is, therefore, no fixed rule of numbers except for those species that always lay one or two eggs.

NESTS ON SEA CLIFFS

GANNET

It is necessary to visit one of the rocky islands off our coast, such as Bass Rock, Ailsa Craig or Grassholm, if the GANNET is to be seen at home. The birds nest in colonies on the ledges of cliffs or the tops of islands and as close as possible in one area. The colonies are sometimes of enormous size, up to 16,000 pairs. The nest itself is made of sea-weed, grass and rubbish, and normally only one egg is laid, having a blue shell covered with a chalky deposit that quickly becomes stained a dirty brown. It is about 3 inches long. Laying begins in April and the bird is single brooded.

Cliff ledges and rocky islets are chosen by the CORMORANT for its nest of sticks and sea-weed lined with grass. This is another species that forms a colony for nesting purposes. The eggs are covered with a thick lime deposit which soon becomes discoloured. The Shag prefers ledges in sea caverns and its eggs, though very like those of the Cormorant, are smaller.

CORMORANT

The FULMAR PETREL also nests in colonies but of varying size. It also chooses steep cliffs and also may nest on the top if undisturbed. Usually no nest is made, the egg being laid on a bare ledge of rock.

The KITTIWAKE builds in colonies on the ledges of precipitous cliffs. The nest is a bulky affair of grass and sea-weed with a neat cup in which the two eggs rest.

KITTIWAKE

Another seabird that breeds in colonies is the GUILLEMOT, often in very large numbers. No nest is made, the single egg being laid on the ledge of a steep cliff. or on the top of an island stack. They vary much in colour from green to white and brown and may be unmarked. The Black Guillemot also makes no nest but it likes the cover of a hole or crevice of steep cliffs in which to lay its two eggs. These are much smaller than those of the Common Guillemot, white or faintly coloured and boldly marked with dark blotches. Several pairs usually nest in close proximity.

The RAZORBILL is rather like the Guillemot in its nesting habits but prefers more broken cliffs. No nest is made, the single egg being laid under cover of a boulder or in a crevice. It is a little smaller than the Guillemot's and usually white, cream or pale brown, more or less heavily marked. These auk-like birds are all single brooded and commence to lay in May or June.

The PUFFIN also nests in colonies but lays its single egg in a burrow taken from a rabbit or made by itself in the soil on the side or top of cliffs. A few feathers and grass are used as a lining for the egg, which is whitish with a few small pale brown or purple spots and about 2⅜ inches long. The Manx Shearwater nests in very similar circumstances, laying one white egg in a burrow which it makes itself.

✕✦✕✦✕✦✕✦✕✦✕✦✕✦✕✦✕✦✕✦✕✦✕✦✕✦✕✦✕✦✕

Some sea-birds lay only one egg. Gulls lay two or three, Pigeons always two, and Plovers normally four. Birds that build cup-shaped nests usually lay three to five, those making a domed nest or building in holes, five to eight, whilst Ducks lay up to a dozen, and the Game-birds even more.
The Guillemot and Razorbill each lay one very pointed egg. The reason given for this is that the egg will only roll on its own axis, and is therefore not so liable to roll off the cliff-ledge on which the bird is in the habit of laying.

NESTS ON CLIFFS

RAVEN

The RAVEN has been mentioned with the crows (page 4) but is included in this grouping because nowadays it so frequently nests on cliffs, both sea and inland. The nest is a large structure of sticks, roots, sea-weed, grass and earth, lined with wool, hair and fine grass.

The eggs of the PEREGRINE FALCON are laid on the bare ledge of a cliff, sometimes inland but more usually by the sea. On occasion the old nest of a Raven or Hooded-crow is known to have been used.

The GOLDEN EAGLE also nests on inaccessible cliffs but makes a great structure of sticks and heather, lined with grass and moss. The eggs, two in number, are white blotched with red-brown and about three inches long. The laying season begins in March and only one brood is reared.

STOCK-DOVES nest in colonies in holes and crannies in cliffs, quarries, old trees and buildings. Practically no nest is made, the two white eggs being laid on the floor of the nest-hole. The Rock-dove, which is much rarer, lays its two white eggs on the ledges of sea caves or in crevices in sea cliffs.

NUMBERS OF EGGS

Certain species lay only one clutch during a season: if this is taken or destroyed it is replaced, but only one brood is reared. This applies to the crows, hawks, ducks, waders, game and sea-birds. Many species, however, rear more than one brood, notably the small perching birds, which mostly manage two or three to four, though there are exceptions even among these, where one brood is the limit.

NESTS IN BUILDINGS

JACKDAW

The JACKDAW, which was mentioned with the crows (page 4) is not particular as to the location of its nest, providing it is some sort of a hole well out of reach from the ground. Buildings are commonly used, chimneys and ruins most often, but quarries, cliffs and old trees are also chosen. The nest is usually a mass of sticks, lined with grass, wool and hair. The Jackdaw builds in colonies.

The STARLING has been included in this section because it so often makes its nest in the chimney, roof or eaves; but a hole in a tree or nest-box is equally suitable. The nest is untidy and made of straw, lined with feathers and other soft materials.

Any hole in a building, rocks, haystack, old tree or in ivy covering them will suit the HOUSE-SPARROW for its nest. Sometimes a domed nest is built in a hedge or tree by this enterprising bird. The nest is nearly always where room permits a rough domed affair of straw and grass, lined with feathers.

The SWALLOW is fond of making its nest on the rafters of a barn or out-building but sometimes the inside of a large chimney is chosen.
Mud carried in the bird's beak and reinforced with small pieces of hay is used to secure a foundation and form a cup, which when dry is lined with feathers and hay.

SWALLOW

HOUSE-
MARTIN

*HOUSE-MARTINS, which nest in colonies, choose
the eaves of buildings under which to construct their
mud nests. Unlike.the Swallow's, that is an open
cup, the top is closed by the eave or similar projection
and an entrance hole is left on one side. It is a
neater affair than the Swallow's and cosier, and it
is lined with feathers. As a general rule a Swallow's
nest is inside a building whereas a Martin's is
outside.*

*The SWIFT finds a cranny under the eaves or a
crevice in the slates where there is just room to lay
its eggs. Nesting material is very scanty, for what
there is, is picked up by the bird on the wing and
cemented in place with saliva.*

*The eggs of the BARN-OWL are laid in a hole or
crevice of a building, even on an unused floor and
often in a hollow tree. Sometimes a box or barrel
is hung up to encourage owls to nest therein because*

*these birds kill large
numbers of rats and
mice. No nesting
material is used,
the eggs being laid
on the open floor.*

BARN-OWL

〉〜〜〜〜〜〜〜〜〜〜〜〜〜〜〜〜〜〜〜〜〜〜〜〜〉

Many wild birds are protected by Law and to take their eggs is a
criminal act; merely to make a collection is wrong. Unless very
carefully looked after, eggs soon become broken, lose their beauty
and charm. If you wish to study them, there are good collections
in museums where you can do so without harming them at all.

There are not enough of the lovely things of life that we can afford
to destroy any, and the nest of a wild bird with its clutch of eggs
intact is one of the most beautiful. So leave them as you find
them, but add your blessing, so you too will be blessed.

Lightning Source UK Ltd.
Milton Keynes UK
UKOW051558020512

191898UK00001B/183/P